FLORA OF TROPICAL EAST AFRICA

BALANITACEAE

Martin J.S. Sands

Deciduous or semi-evergreen spiny trees or shrubs. Spines simple or branching, derived from the distal of 2 or more buds, axillary or at a varying supra-axillary interval[1]. Leaves alternate or spirally arranged, either reduced to simple, minute and usually triangular, mostly caducous, exstipulate scale-leaves, or bifoliolate, sessile or petiolate, with a linear or triangular, sometimes persistent, terminal foliole[2]; stipules caducous or persistent, triangular to filiform, entire. Flowers actinomorphic, sepals and petals free, bisexual, 4–5-merous, rarely solitary, usually in a 2–many-flowered, bracteate, mostly cymose and subumbellate to fasciculate inflorescence, borne at spineless or spiniferous nodes on the stem or spines. Sepals imbricate, often reflexed and caducous, one or both margins glabrous, silky-hairy within. Petals spreading, sometimes reflexed, usually irregular, bent or contorted at the apex, yellow to green, glabrous outside, glabrous or hairy within. Stamens 8–10, free, glabrous; anthers dorsifixed, 2-thecous, dehiscing by longitudinal slits. Disc intrastaminal, annular, succulent, corrugated and minutely papillose, surrounding the base of the ovary. Ovary superior, hemispherical to ovoid, 4–5-locular, each locule containing 1 pendulous ovule; style simple, terminal, glabrous, stigma small and simple. Fruit a 1-seeded drupe on a thickened pedicel; exocarp leathery or brittle, mesocarp variable from pulpy to fibrous and oily, endocarp hard, often woody, splitting distally at germination; seed with plano-convex cotyledons.

Monotypic.

For the purpose of this Flora, *Balanitaceae* is recognized as an independent family. The genus *Balanites* has been variously classified and its supra-generic position continues to be debated. Recent molecular work by Sheahan & Chase (Syst. Bot. 25(2): 380 (2000)) has shown that *Balanites* is embedded in the Tribuloid clade within *Zygophyllaceae*, the Angiosperm Phylogeny Group (Ann. Missouri Bot. Gard., 85: 531–553 (1998)) also include *Balanites* in *Zygophyllaceae*.

BALANITES

Delile, Descr. Egypte, Hist. Nat. 2: 221, t. 28, 1 (1813); Sands in K.B. 56: 1–128 (2001)

Description as for the family.

Nine species, two species in India and Burma, the others distributed throughout much of Africa, with one also occurring in Yemen and another extending into Arabia and the Jordan valley.

Several species of *Balanites* are of economic importance in Africa. A major use is that of the fruit of several species, especially the 'desert date' (*B. aegyptiaca*) which is edible and yields a valuable oil as well as the pharmaceutically important steroid, diosgenin, and a saponin-glucoside toxic to cold-blooded animals that are hosts to guinea-worm and stages of bilharzia. The leaves and inflorescences have several uses including as vegetables. The wood of most species has many uses such as furniture making and turnery, and bark and roots especially are variously used in traditional medicine. All aerial parts are browsed by livestock.

---

[1] distance above the subtending leaf
[2] vestigial terminal leaflet

1

Key to flowering material

Balanites sp. '**Omo Valley**' is known only from sterile material and therefore not included in this key.

1. Flowers 5-merous; petals hairy within ·················· 2
   Flowers 4- or 5-merous; petals glabrous within ·················· 3
2. Tree usually more than 30 m high; spines forked or branching several times, the spine-branches usually more or less equal; fruit 8–12 cm long ············ **1. B. wilsoniana**
   Shrub or tree less than 20 m (rarely 25 m) high; spines simple or branching with the spine-branches usually unequal; fruit usually less than 6 cm long, rarely to 8 cm ············ **2. B. maughamii**
3. Spines not bearing flowers or leaves; ovary elongating early in development ·················· **3. B. aegyptiaca**
   Spines bearing flowers or leaves; ovary not or scarcely elongating in early development ·················· 4
4. Flowers 5-merous ·················· **4. B. pedicellaris**
   Flowers 4-merous ·················· 5
5. Spines arising (1–)2–5(–10) cm above the subtending leaf axil; secondary spines present; ovary hairy or glabrous ·· **5. B. rotundifolia**
   Spines arising 0.5–2 cm above the subtending leaf axil; spines sometimes branching but secondary spines absent; ovary glabrous ·················· **6. B. glabra**

Key to sterile specimens

This key assumes the presence of spines or at least a field report of their occurrence. *Spinules* are small secondary spines borne on the primary spine and usually at a wide angle to it.

1. Spines naked without leaves or flowers or their scars ·················· 2
   Spines bearing leaves or exhibiting leaf and/or flower or inflorescence scars ·················· 5
2. Small shrub less than 0.5 m high; spinules present; leaflets sharply acute or apiculate ·················· **7. B. sp.'Omo Valley'**
   Large shrub or small tree; spinules absent, leaflets obtuse, rounded or bluntly acute to acuminate ·················· 3
3. Spines usually forked or branching with deflection of the primary spine, at least on the trunk and main branches ·················· 4
   All spines simple, unbranched, or branch-spines rare, and short clearly subordinate without deflection of the primary spine ·················· **3. B. aegyptiaca**
4. Large tree usually more than 30 m high; forked or branching spines mostly borne on the trunk and branches, infrequent and simple on branchlets; remains or scars of inflorescences supra-axillary, up to 5(–10) mm above the axil; stipules caducous; indumentum of petioles and young growth silvery-grey ·················· **1. B. wilsoniana**
   Tree rarely exceeding 20 m high; spines on younger stems as well as on the upper trunk and branches; remains or scars of inflorescences axillary; stipules usually persistent; indumentum of petioles and young growth yellowish- or greyish-green to buff ·················· **2. B. maughamii**

5. Spines sometimes branching, tending to widen dorsiventrally at the base, 0.5–3 cm above the axil; spinules absent; foliage-leaves infrequent; scale-leaves occurring on the parent stem, sometimes persistent · · · · · · · · · · · · · · · · · 6. *B. glabra*
   Spines frequently branching, scarcely widening at the base, 0.5–10 cm above the axil; spinules often present; foliage-leaves frequent; scale-leaves only evident on growth or secondary spines · 6
6. Primary spine and branch-spine/spinule supra-axillary intervals 0.5–1.2(–2.6) cm and 1–2 mm respectively; leaflets broadly obovate to spathulate, cuneate; scale-leaves occurring on the spinules, or sometimes on growth, up to 1.3 mm long; stipules caducous · · · · · · · 4. *B. pedicellaris*
   Primary spine and branch-spine/spinule supra-axillary intervals (1–)2–5(–10) cm and (2–)4–7 mm respectively; leaflets usually orbicular to broadly obovate, rounded or broadly cuneate; scale-leaves occasional on young growth, 2–3 mm long; stipules often persistent · · · · · · 5. *B. rotundifolia*

1. **B. wilsoniana** *Dawe & Sprague* in Dawe, Bot. Miss. Uganda Prot.: 14, 23 & 40 (April 1906) & (desc. expans.) in J.L.S., Bot. 37: 506 (Nov. 1906); Z.A.E. 2: 422 & t. 47 (1910); I.T.U.: 230 (1940) & ed. 2, with Dale: 407 (1952); Gilbert in F.C.B. 7: 66 (1958) pro parte; Hamilton, Uganda For. Trees: 194 (1981); Sands in K.B. 56: 20, t. 5, map 1 (2001). Type: Uganda, Toro District, Kibale Forest, *Dawe* 511 (K!, holo.)

Usually evergreen tree 30–45(–50) m high with an irregular, sometimes open crown; trunk up to 1.2 m in diameter, sometimes buttressed and usually very deeply fluted; bark smooth to irregularly rough and warty, yellowish to light brown, sometimes with scattered small black spheres of hardened resinous exudate; branchlets frequently from swollen nodes, often deciduous leaving a conical pit, glabrous or puberulous and glabrescent, yellowish-green occasionally becoming blackened. Spines borne at a wide angle on the trunk and branches, up to 15 cm long and 9 mm diameter at the base, without leaves or flowers, forked or branching several times, the branch-spines subtended by scale-leaves when young; spines on flowering and some sterile branchlets infrequent, usually simple and short, up to 1(–2) cm long borne 5–12 mm above the axil; spinules absent. Leaves on sterile and fertile branchlets, usually persistent; stipules narrowly triangular, 0.2–0.5 mm long, soon caducous; petiole 1–3.1 cm long; petiolules 0.3–1.8 cm long; leaflets ovate to ovate-elliptic with the inner half often slightly smaller than the outer, 5.8–11.5 cm long, 2.7–7.5 cm wide, acute to acuminate, often developing a drip-tip 1.5–3 cm long on sterile shoots, more or less equally rounded or abruptly cuneate, the two sides sometimes joining the petiole 1–2 mm apart, membranous, eventually coriaceous; venation often prominent, glabrous, but sparsely puberulous on the main vein beneath at first; foliole 0.3–3(–14) mm long, sometimes early caducous. Inflorescence 2–5(–8)-flowered, subfasciculate or condensed, often an umbellate cyme, 0.5–1 cm from the leaf-axil on stems of the current year, or several on an axillary, usually leafless shoot, grey-puberulous to tomentellous; peduncle absent or up to 18 mm long. Flowers 5-merous; sepals elliptic to ovate, remaining concave, 3–6 mm long, 1.5–3.7 mm wide, acute, usually caducous as the bud opens, shortly grey-puberulous to tomentellous outside, the glabrous margin narrow; petals pale yellowish green, oblong-elliptic to oblanceolate, becoming reflexed at anthesis, 6–10 mm long, 1.5–2.5 mm wide, acute, the glabrous tip sometimes slightly contorted, villous within; stamens 10, spreading-erect; anthers 1–1.5 mm long, 0.5–1 mm wide; ovary 1–2 mm high, densely white- to grey-hairy; style 1–1.5 mm long. Fruit only one developing from an inflorescence, soon elongating in early development, brown to yellow at maturity, ovoid or ellipsoid to broadly fusiform, (6.5–)8–12 cm long, (3.5–)5.8–7 cm wide, tapering to obtuse ends, glabrous and dark, pulpy, oily and malodorous within; seed loose at maturity, cream, ellipsoid to fusiform, up to 4.3 cm long.

var. **wilsoniana**; Sands in K.B. 56: 25, t. 5/a–k, map 1 (2001)

Pedicels up to 1.3 cm long at anthesis. Petals villous within, not exceeding 7 mm long.

UGANDA. Bunyoro District: Budongo Forest, 1935, *Eggeling* 1634! & 15 May 1941, *Thomas* 3894!; Mengo District: Mabira Forest, Mulange, Jan.–Feb. 1920, *Dummer* 4396!
DISTR. U 2, 4; from Ivory Coast in the west to Uganda and southwards into Congo (Kinshasa)
HAB. Moist forest; 1100–1200 m

NOTES. A var. *mayumbensis* occurs in Congo (Brazzaville) and Angola, and var. *glabripetala* in Nigeria. They are distinct in the longer petals and the longer pedicels at anthesis; var *glabripetala* has glabrous petals..
Many specimens of *B. maughamii* subsp. *acuta* from Kenya and Tanzania were formerly identified as *B. wilsoniana*.

2. **B. maughamii** *Sprague* in K.B. 1913: 136, 138 (1913); V.E. 3: 743 (1915); E.&P. Pf. ed. 2, 19a: 182 (1931); T.T.C.L.: 571 (1949); Launert in F.Z. 2: 221, t. 42 (1963); Sands in K.B. 56: 29, t. 6, map 2 (2001). Types: Mozambique, Manica e Sofala Division, Madanda Forest, *Dawe* 428 (K!, lecto., BM!, isolecto.; chosen by Sands)

Deciduous or semi-deciduous tree up to 20(–25) m high, with a spreading, rounded crown, rarely a low shrub 1.5–2 m high; trunk to 1.3 m in diameter, fluted; bark usually smooth, yellowish-brown, mottled or grey, but sometimes becoming rough, splitting longitudinally dividing into small, dark grey sections; branchlets and spines pubescent or finely puberulous, eventually glabrescent, the branchlets yellowish- to greyish-green, often becoming black on one side or in patches in the second year, sometimes deciduous leaving a swollen, woody scar. Spines axillary or rarely up to 5 mm above the axil or on older wood, (1–)3–6(–15) cm long, spines on flowering shoots usually absent or simple and 3–5 mm long, otherwise frequently branched, occasionally more than once, often appearing forked. Leaves with stipules persistent, triangular, 1–3 mm long, 0.8–1.5 mm wide; petiole 0.8–5 cm long; petiolules 0.1–2.5 cm long; leaflets elliptic to broadly ovate or suborbicular, 2.4–12 cm long, 1.5–9.7 cm wide, usually asymmetrical, apex rounded, obtuse, acute to apiculate or shortly acuminate, base rounded or broadly cuneate, becoming coriaceous, glabrous or variably pubescent on both surfaces, eventually glabrescent; foliole 2–3 mm long, usually linear, eventually caducous. Inflorescence a (1–)3–7-flowered subfasciculate or umbellate cyme, 1–2 in a leaf-axil of 1-year old stems or several on an axillary, often leafy, shoot, variably tomentellous to pubescent or puberulous, indumentum yellowish-green to buff; peduncle absent or up to 1 cm long; pedicels 0.3–1.3 cm long. Flowers 5-merous, often scented, sepals and petals reflexed after anthesis; sepals yellowish-green, ovate to obovate, 4–5.6 mm long, 2–3.2 mm wide, acute or rounded, tardily caducous, pubescent or densely puberulous outside, the glabrous margin broad; petals green or greenish-yellow, oblong-lanceolate to oblanceolate, 5–9 mm long, 1.5–2.5 mm wide, frequently with a contorted and glabrous tip, villous within; stamens 10, spreading, erect, anthers 1.2–1.9 mm long; ovary densely and stiffly hairy, 1 mm high; style 0.5–1.5 mm long. Fruit, usually one developing from an inflorescence, pale to reddish-brown when mature, oblong-ellipsoid to cylindrical and depressed at both ends or ovoid, 3.5–8 cm long, 2.2–3.7 cm wide, obtuse apically, often with 5 shallow grooves, elongating in early development, glabrous, eventually with a brittle surface, spongy and fibrous, dark and oily within with a hard cream-coloured endocarp; seed ellipsoid to fusiform, grooved, cream, up to 2.5 cm long, becoming loose in the mature fruit.

DISTR. (of species as a whole) Kenya, Tanzania, Zambia, Malawi, Mozambique, Zimbabwe, Swaziland, South Africa

NOTE. *B. maughamii* is undoubtedly very variable in a number of characters and, even on a single tree, sterile and fertile shoots often exhibit marked differences in leaflet shape and size and in the density of the indumentum.

Leaflets pubescent, usually densely so, or at least hairy below on the mid-vein; apex rounded or obtuse on fertile shoots · · · · · · · · subsp. *maughamii*
Leaflets glabrous on both surfaces, even when young; apex acute or shortly acuminate on fertile shoots · · · · · · · · subsp. *acuta*

subsp. **maughamii**; Sands in K.B. 56: 35, t. 6/n–k, map 2 (2001)

Leaflets pubescent above and below or, rarely, at least hairy on the mid-vein beneath, those of leaves on the fertile shoots obtuse or rounded at the apex or if acute then the tip blunt. Indumentum of the inflorescence usually densely pubescent to tomentellous.

TANZANIA. Songea District: 1.5 km E of Songea, 13 Feb. 1956, *Milne-Redhead & Taylor* 8683!; Lindi District: Lake Lutamba, 40 km W of Lindi, 20 Sept. 1934, *Schlieben* 5364! & Rondo escarpment, approach to Lake Lutamba, Nov. 1953, *Eggeling* 6716!
DISTR. **T** 8; Zambia, Malawi, Mozambique, Zimbabwe, Swaziland, South Africa
HAB. Dry bushland or *Brachystegia-Uapaca* woodland; 250–1100 m

SYN. *Balanites sp.* of Oliv., F.T.A. 1: 315 (1868); Sprague in J.L.S., Bot. 37: 507 (1906); Sim, For. Fl. Port. E. Afr.: 25 (1909)
[*Trachylobium mossambicensis* sensu Sim, For. Fl. Port. E. Afr.: 51, pl. 56 (1909), pro parte; Mello Géraldès, J. Agric. Trop.: 12, 233 (1912) (as *T. 'moçambicensis'*), *non* Klotzsch]
Balanites dawei Sprague in K.B. 1913: 137, 140 (1913); V.E. 3: 743 (1915); E.&P. Pf. 19a: 182 (1931); T.T.C.L.: 571 (1949). Type: Mozambique, Manica e Sofala Division, Madanda Forest, *Dawe* 435 (K!, holo.)
B. sp. 1 of White, For. Fl. N. Rhodesia: 168 (1962)

NOTE. Intermediates between the two subspecies occur from southern Tanzania southwards. In herbaria, sterile and fertile collections of subsp. *maughamii* may look very different, but sheets which display both fertile and sterile shoots clearly show that they are the same taxon.

subsp. **acuta** Sands in K.B. 56: 37, t. 4/Bv, t. 6, map (2001). Type: Kenya, Kwale District, Jadini [Jardini], *Greenway* 9803 (K!, holo.; EA, iso.)

Lamina of the leaflets completely glabrous (except very rarely on the lower mid-vein below) and acute or acuminate on flowering as well as sterile shoots. The indumentum of the inflorescence also differs from subsp. *maughamii* in being more even, consisting of much shorter hairs.

KENYA. Machakos District: Chyulu foothills, May 1938, *Bally* 636!; Kwale District: Diani beach near Mombasa, 9 Aug. 1958, *Willan* 337!; Kilifi District: Manyani, just south of Malindi Point, 2 April 1981, *Gilbert* 6038!
TANZANIA. Tanga District: Kigombe beach, 11 km NE of Pangani, 12 July 1953, *Drummond & Hemsley* 3272!; Handeni District: Kwamarukanga, 15 Dec. 1971, *Shabani* 797!; Kilwa District: Selous Game Reserve, Kingupira Forest, 9 July 1975, *Vollesen* 2546 !
DISTR. **K** 4, 7; **T** 3, 6, 8; not known elsewhere
HAB. Evergreen coastal forest, coastal or riverine thicket, groundwater forest; 0–1000 m

SYN. [*Balanites wilsoniana* sensu Hutchinson & Dalziel, F.W.T.A. 1: 484 (1928) pro parte quoad loc. Kenya; Dale in T.S.K. ed. 2: 101 (1936) & Woody Veg. Coast Prov. Kenya (Imp. For. Inst. Pap. 18): 9 (1939); T.T.C.L.: 572 (1949) pro parte excl. descr. fruit; K.T.S.: 534 (1961) pro parte; K.T.S.L.: 378, map (1994), *non* Dawe & Sprague].

3. **B. aegyptiaca** (*L.*) *Delile*, Descr. Égypte, Hist. Nat.: 221 & t. 28, fig. 1 (1813); Oliv., F.T.A. 1: 315 (1868); E.&P. Pf. 3, 4: 355 (1896) pro parte; E.P.A.: 365 (1956) pro parte; Gilbert in F.C.B. 7: 66 (1958); Launert in F.Z. 2: 223 (1963); K.T.S.L.: 377, fig., map (1994); Sands in K.B. 56: 51, t. 9 & 11 map 4 & 6 (2001). Type: t. 39 (excl. fruits) in Alpinio, Pl. Aegypt. Liber, 38 (1640) as Agialid, lectotype, *fide* Basak in Fasc. Fl. Ind. 4: 20 (1980) pro parte. Epitype: Tanzania, Tanga District, Mombe Forest Reserve, *Semsei* 3580 (K!, epitype, chosen by Sands)

A semi-evergreen or sometimes deciduous shrub or small tree up to 12(–15) m high, usually spiny, extremely variable in many of its characters; bole usually straight, to 60 cm diameter, often fluted, branches spreading irregularly or pendulous, sometimes forming a rounded crown; bark hard, becoming rough, corky and deeply fissured, dark grey; branchlets and spines greyish green becoming light brown, at first minutely puberulous to tomentellous, usually glabrescent. Spines borne at wide angle to the parent stem, axillary or to 4 mm above the axil, straight or slightly curved, (0.4–)2–8(–11.5) cm long, 2–6 mm diameter at the base, terete with a sharp tip or subulate, without leaves or flowers, only very rarely bearing a short branchspine. Leaves subsessile or with a petiole up to 3.5 cm long; stipules 0.5–1 mm long, triangular, caducous, finely puberulous, glabrescent; leaflets very variable, narrowly spathulate or elliptic to broadly ovate or obovate, sometimes very broadly spathulate or almost orbicular and eccentric, 0.6–6.8 cm long, 0.3–5 cm wide, sessile or with petiolules 0.1–1 cm long, thin to coriaceous, apex bluntly acute to obtuse or rounded, occasionally emarginate, cuneate or narrowly decurrent at the base, closely and minutely puberulous, often glabrescent, or sometimes densely tomentellous or tomentose; foliole 1–2.5 mm long, often caducous, linear and bent. Inflorescence of (1–)2–20 or more flowers variously arranged in cymose fasicles at spineless or spiniferous nodes, or closely arranged on shoots of short internodes, sometimes more or less spiciform, finely tomentellous to tomentose; peduncle, if present, 2–8 mm long; pedicels 0.4–2 cm long. Flowers 5-merous, sweet-scented; sepals spreading to reflexed, ovate, 3.5–5.5 mm long, 1.5–2.5 mm wide, acute to acuminate, tomentellous to tomentose on the outside, sometimes early caducous; petals spreading, sometimes reflexed, green to pale yellow or creamy white, lanceolate or narrowly elliptic to obovate-oblong, 4.5–7 mm long, 1–2.5 mm wide, rounded or obtuse to acute, sometimes contorted at the apex, glabrous within; stamens 10, spreading; anthers 0.7–1.8 mm long, 0.4–0.7 mm wide; ovary 0.7–1 mm high, puberulous or pubescent to tomentose; style 1.5–3.5 mm long. Fruit elongating markedly in very early development, very variable when mature, ripening greenish brown to yellow, subspherical or ovoid to ellipsoidal, 2.3–4.7 cm long, 1.3–2.8 cm wide, rounded, truncate or sometimes sulcate at both ends or apex tapering, conical, leathery at first, becoming brittle on the outside, inside yellowish brown, fibrous and oily, surrounding a hard, woody endocarp containing the seed.

DISTR. (of species as a whole) throughout much of Africa from Mauritania and Senegal to Somalia and from Egypt southwards to Zambia and Zimbabwe; W Arabia and the Jordan valley

1. Leaflet indumentum conspicuous and/or dense · · · · · · · ·   c. var. *tomentosa*
   Leaflet indumentum sparse and/or early glabrescent or
   minutely puberulous · 2
2. Petiole 0–7 mm long; inflorescence 1–6-flowered usually at
   spineless nodes ·   a. var. *aegyptiaca*
   Petiole 7–37 mm long; inflorescence 6–11-flowered, nearly
   always at spinous nodes ·   b. var. *quarrei*

a. var. **aegyptiaca;** Sands in K.B. 56: 56, t. 9, map 4 (2001)

Leaflet indumentum sparse and/or early glabrescent or minutely puberulous; petiole 0–7 mm long. Inflorescence 1–6-flowered usually at spineless nodes. Fig. 1 (page 7).

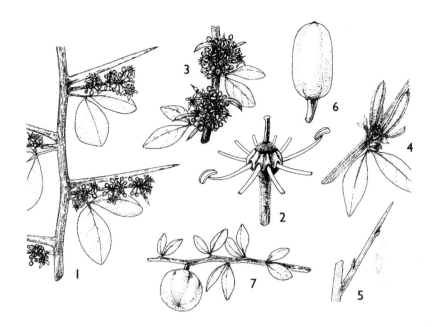

FIG. 1. *BALANITES AEGYPTIACA* var. *AEGYPTIACA*. **1**, stem with stout spines and flowers on condensed shoots of short internodes, × 0.5; **2**, flower with petals and sepals removed, × 4; **3**, young infrutescence, × 0.5; **4**, developing fruits, × 0.5; **5**, young spine bearing scale leaves, × 2; **6**, mature fruit, × 0.5; **7**, stem with sessile leaves, narrow leaflets and rounded fruits, × 0.5. 1 from *Jacques-Felix* 3406, 2, 3 from *van Someren* 5915, 4 from *de Wilde & Amshoff* 6543, 5 from *Mooney* 5754, 6 from *Chaffey* 621, 7 from *Grazorsky* 4. Drawn by Margaret Tebbs.

UGANDA. Acholi District: Paimol hill, 27 Dec. 1947, *Dawkins* 309! & SE Imatongs, Agoro, 8 April 1945, *Greenway & Hummel* 7325!; Teso District: Serere, Feb. 1933, *Chandler* 1110!
KENYA. Northern Frontier District: Moyale, 21 April 1952, *Gillett* 12874!; Turkana District: Napau Pass, 21 Feb. 1953, *Dawkins* 791!; Nairobi/Masai District: Athi plains, 14 March 1934, *van Someren* 5915!
TANZANIA. Masai District: Longido Mt, 14 Jan. 1936, *Greenway* 4364!; Handeni District: Kwafumbili, Nov. 1949, *Semsei* in *F.H.* 2921!; Morogoro District: near Morogoro, 4 Sept. 1930, *Greenway* 2529!
DISTR. **U** 1–4; **K** 1–6; **T** 1–6; throughout much of Africa from Mauritania to Somalia and from Egypt southwards to Zambia and Zimbabwe; W Arabia and the Jordan valley
HAB. Wooded and scattered tree grassland, deciduous bushland, especially where water table is seasonally high; 350–1800(–2100) m

SYN. *Ximenia aegyptiaca* L., Sp. Pl.: 1194 (1753)
 Agialida aegyptiaca (L.) Adans., Fam. Pl. 2: 508 (1763) (as *Agialid aegyptiaca*); Kuntze, Revis. Gen. Pl.: 103 (1891)
  A. membranacea Tiegh. in Ann. Sci. Nat., ser. 9, 4: 228 (1906). Type: Sudan, Dongola Province, Dabbah, *Ehrenberg* s.n. ('1820–1826') (P, holo.)
  A. abyssinica Tiegh. in Ann. Sci. Nat., ser. 9, 4: 229 (1906). Type: Ethiopia, 1200 m, *Schimper* 1222 (P, holo., K!, iso.)
  A. cuneifolia Tiegh. in Ann. Sci. Nat., ser. 9, 4: 229 (1906). Type: Ethiopia, Dscha-Dscha, Gurrserfa, 20 Dec. 1854, *Schimper* s.n. (P, holo.)
  A. latifolia Tiegh. in Ann. Sci. Nat., ser. 9, 4: 230 (1906). Type: Ethiopia, Shoa (Shewa), 1845, *Rochet d'Hericourt* s.n. (P, holo.)

A. *nigra* Tiegh. in Ann. Sci. Nat., ser. 9, 4: 230 (1906). Type: Ethiopia, Shoa (Shewa), 1845, *Rochet d'Hericourt* s.n. (P, holo.)

A. *schimperi* Tiegh. in Ann. Sci. Nat., ser. 9, 4: 230 (1906). Type: Ethiopia, Tacazzé Mts, 1800 m, *Schimper* 1222 (P, holo., K!, iso.)

Balanites fischeri Mildbr. & Schltr. in E.J. 51: 157 (1913). Type: Tanzania, Shinyanga District, Usiha, *Fischer* 123 (?B†, holo.)

B. latifolia (Tiegh.) Chiov., Fl. Somal. 2: 46, fig. 5 (1932)

B. suckertii Chiov. in Atti Soc. Naturalisti Mat. Modena 66: 3 (1935). Type: Somalia, near Uanle-Uen, *Suckert* 73 (FT, holo., K!, iso.)

NOTE. Apparently far less common than var. *tomentosa* in central Tanzania.

b. var. **quarrei** (*De Wild.*) *G. Gilbert* in F.C.B. 7: 68, t. 10 (1958); Schmitz in B.J.B.B. 96: 328 (1963); Sands in K.B. 56: 69, t. 11/a, b, map 6 (2001). Type: Congo (Kinshasa), Shaba, Elakat, *Quarré* 1415 (BR!, holo.)

Petiole to 3.7 cm long; leaflets rounded to broadly cuneate at the base, sparsely hairy, glabrescent; spines stout, to 11 cm long. Inflorescence usually arising at a spinous node on old stout stems.

TANZANIA. Ufipa District: Namanyere–Karonga road km 3, 4 March 1994, *Bidgood et al.* 2599!
DISTR. **T** 4; SE Congo (Kinshasa), Zambia
HAB. "On termite mound in grassland"; 1550 m

SYN. *B. quarrei* De Wild. in Contrib. Fl. Katanga, suppl. 5: 33 (1933)

c. var. **tomentosa** (*Mildbr. & Schltr.*) *Sands* in K.B. 56: 72, t. 11/c–e, map 6 (2001). Type: Tanzania, without precise locality, *Fischer* 125 (B†, holo.); Singida District: near Matalele, near Ipusuro, *B.D. Burtt* 723 (K!, neo., BM!, EA!, isoneo., chosen by Sands)

Petiole to 1 cm long; spines to 1.5(–3) cm long. Pedicel and calyx tomentose. Flowers usually at spinous nodes.

TANZANIA. Shinyanga District: Shinyanga, [rec.] Nov. 1938, *Koritschoner* 1701!; Kondoa District: Kolo hills, 18 June 1973, *Ruffo* 729!; Mpwapwa District: Kiboriani Mt, 5 July 1938, *Hornby* 910!
DISTR. **T** 1, 5; not known elsewhere
HAB. Deciduous thicket and bushland, *Brachystegia* woodland; 900–1450 m

SYN. *B. tomentosa* Mildbr. & Schltr. in E.J. 51: 159, fig. 1f (1913); V.E.3: 743, fig. 34f (1915); E.&P. Pf. 2, 19a: 182, fig. 87f (1931); T.T.C.L.: 572 (1949)

NOTE. This variety is at first sight very distinctive with its short spines and a conspicuously dense indumentum. However, some material of other varieties of *B. aegyptiaca* is equally hairy, and short spines also occur, including those on intermediate specimens from adjacent areas such as Tabora District, and on some specimens of var. *quarrei*. Collections from Tabora (**T** 4, *Lindeman* 8 & 603) are short-spined, but are here excluded from var. *tomentosa* and referred to *B. aegyptiaca* var. *aegyptiaca*. They have larger, more glabrescent leaflets and longer petioles and there appears to be a link with var. *quarrei* and a 'southern form' of *B. aegyptiaca*, occurring south of the Tanzanian border. Var. *ferox* and var *pallida* occur respectively in W Africa and around the Gulf of Aden.

4. **B. pedicellaris** *Mildbr. & Schltr.* in E.J. 51: 162 (1913); V.E. 3: 743, fig. 348d (1915) & in E.&P. Pf. 2, 19a: 182, fig. 87d (1931); I.T.U.: 230 (1940) & ed. 2, with Dale: 407 (1952); T.T.C.L.: 571 (1949); K.T.S.: 534 (1961); Launert in F.Z. 2: 223 (1963); Launert & Paiva in Fl. Moçamb. 39: 3 (1969); Sands in Fl. Ethiop. 3: 436 (1990); K.T.S.L.: 378, map (1994); Sands & Thulin in Fl. Somalia 2: 171 (1999); Sands in K.B. 56: 74, t. 12, map 7 (2001). Type: Kenya, Teita District: Voi, *Mildbraed* 19 (B†, holo., K!, PRE!, iso.)

A spiny deciduous shrub or small tree up to 8 m high, multistemmed or with a fluted trunk to 30 cm diameter; branches often spreading and then pendulous to form a rounded crown; bark smooth, sometimes papery and peeling, dark grey, becoming rough or corrugated and cracking into small lenticular sections;

branchlets and spines yellowish green to grey-green, becoming reddish brown and eventually grey, at first densely tomentellous, glabrescent. Primary spines borne at a wide angle to the parent stem, 5–26 mm above the axil, often slightly bent at the nodes, 4.6–14.5 cm long, 2–3(–5) mm diameter at the base, frequently bearing branch-spines (themselves spinuliferous or bearing non-spiniform secondary shoots). Leaves on the stems, spines and secondary branches; stipules 2–2.5 mm long, narrowly triangular, tomentose, early caducous; petiole 1–9 mm long; petiolules absent or up to 3.5 mm long; leaflets often greyish green, broadly obovate to spathulate, 0.3–4.6 cm long, 0.3–3 cm wide, apex rounded or obtuse to emarginate, sometimes shortly apiculate, base cuneate, thinly coriaceous to thick and sub-succulent, dull, puberulous or tomentellous to glabrescent, the veins usually obscure; foliole 1–2 mm long, eventually caducous. Inflorescence axillary on the main axis or from spine-buds, the flowers solitary or 2–4(–8) in a fascicle or sometimes in a shortly pedunculate, subumbellate cyme, variably tomentellous to grey-pubescent; peduncle absent or up to 1 cm long; pedicels 0.5–2.4 cm long. Flowers 5-merous, often sweet-scented, the buds apiculate, rarely rounded; sepals ovate, 3.5–8 mm long, 2–3 mm wide, acute to acuminate, tomentellous outside, the glabrous margin very narrow, caducous; petals spreading or reflexed, green or greenish-yellow, elliptic to elliptic-lanceolate, 6–9 mm long, 2–3.4 mm wide, sometimes bent at the tip, glabrous within; stamens 10, spreading or spreading-erect; anthers 1.5–2.8 mm long, 0.5–1 mm wide; ovary ± 1.5 mm high but obscured by dense white hairs protruding from the disc before the ovary develops; style 1–4 mm long, usually persisting in fruit. Fruit elongating only slightly in early development, enlarging evenly to 1.2–4 cm long, 1.3–2.2 cm diameter, becoming ovoid or subglobose, ripening to yellow and then dull orange, often shallowly depressed at both ends, at first finely puberulous, usually glabrescent, thin and usually brittle on the outside, spongy and fibrous within with a thin hard endocarp; seed ellipsoid, cream, up to 1.5 cm long, becoming loose in the mature fruit.

Leaflets 1.2–4.6 cm long; pedicels at anthesis (0.6–)0.8–2.4 cm long · subsp. *pedicellaris*
Leaflets 0.3–1.4(–2.5) cm long; pedicels at anthesis 0.5–0.8 (–1) cm long · subsp. *somalensis*

subsp. **pedicellaris**; Sands in K.B. 56: 78, t. 12/a–k, map 7 (2001)

Shrub or small tree to 8 m; leaflets usually more than 1.2 cm long. Pedicels usually more than 0.8 cm long at anthesis. Fig. 2/1 (page 11).

UGANDA. Karamoja District: Rupa, March 1958, *J. Wilson* 406! & Moroto R., no date, *Brasnett* 80!
KENYA. Turkana District: Elamach, 8 km SSE of Lokichar, 19 June 1970, *Mathew* 6860!; Voi District: Tsavo National Park East, Voi Gate, 5 Jan. 1967, *Greenway & Kanuri* 12971!; Kwale District: between Samburu & Mackinnon Road, 2 Nov. 1953, *Drummond & Hemsley* 4109!
TANZANIA. Lushoto District: 8 km SE of Mkomazi, 2 May 1953, *Drummond & Hemsley* 2389! & W Usambaras, Mkomazi, 7 Sept. 1935, *Greenway* 4058!; Iringa District: 10 km W of Mtera Bridge, 13 Aug. 1970, *Thulin & Mhoro* 718!
DISTR. U 1; K 1–3, 7; T 2, 3, 5–7; Malawi, Mozambique, Zimbabwe, Botswana, Swaziland and South Africa
HAB. Deciduous bushland, thicket, woodland and wooded grassland, often near streams; 0–1450 m

SYN. *B. horrida* Mildbr. & Schltr. in E.J. 51: 160 (1913); V.E. 3: 743, fig. 348c (1915); T.T.C.L.: 571 (1949). Type: Tanzania, Kilosa District, near Kilosa, *Busse* 170A (B†, holo., K!, iso.)

NOTE. Two specimens of *B. pedicellaris* from the Tana River area near Ijara (K 1) are similar to some specimens of *B. rotundifolia*. While superficially appearing to match *Bally* 2076 and 2116, *Bally* 2091 (*B. rotundifolia* var. *setulifera*, collected in the same area), exhibits greater separation of spines from the nodes, remains of a 4-merous disc at the base of the fruit and short stiff hairs on a few of the leaflets.

subsp. **somalensis** (*Mildbr. & Schltr.*) *Sands* in K.B. 38: 40 (1983) & in Fl. Ethiop. 3: 436, fig. 121.1, 13–15 (1990); Sands & Thulin in Fl. Somal. 2: 171, Fig. 111/m–o (1999); Sands in K.B. 56: 81, t. 12L, map 7 (2001). Type: Ethiopia, Webi Mana, *Ellenbeck* 1987 (B†, holo.); Somalia, near Erbah, *Ruspoli & Riva* 1026 (FT!, neo., chosen by Sands)

Densely spinous shrub or occasionally a small tree otherwise characterised by its small, often densely tomentellous leaflets, very rarely exceeding 1.4 cm long, correlated with pedicels short at anthesis, usually 8 mm long or less.

KENYA. Northern Frontier Province: Mandera, 24 May 1952, *Gillett* 13306!
DISTR. **K** 1; Ethiopia, Somalia
HAB. Deciduous bushland; ± 500 m

SYN. *B. somalensis* Mildbr. & Schltr. in E.J. 51: 159 (1913); V.E. 3: 743, fig. 348e (1915); E.&P. Pf. 2, 19a: 182, fig. 87e (1931); E.P.A.: 366 (1956)
  B. somalensis Mildbr. & Schltr. var. *cinereo-corticata* Chiov., Res. Sci. Somalia Ital.: 39 (1916); E.P.A.: 367 (1956). Type: Somalia, Matagassile on Juba R. bank, *Paoli & Stefanini* 834 (FT!, holo.)

5. **B. rotundifolia** (*Tiegh.*) *Blatt.* in Rec. Bot. Survey India, 8: 109 (1919); Sands in K.B. 38: 40 (1983) & in Fl. Ethiop. 3: 434, fig. 121.1, 10–12 (1990); K.T.S.L.: 378, map (1994); Sands & Thulin in Fl. Somal. 2: 171, Fig 111/j–l (1999); Sands in K.B. 56: 90, t. 14, map 9 (2001). Types: Aden, 1860, *Courbon* s.n. (P!, syn.), and Djibouti, Gulf of Tadjourah, Obock, 1886, *Faurot* s.n. (P!, syn.)

A spiny evergreen shrub or small tree up to 6(–8) m high (frequently much shorter), with a low bushy habit, densely branched, or with a trunk up to 40 cm diameter; bark grey-brown, strongly fissured; sap gummy; branchlets grey-green or yellowish-brown and, like the young spines, glabrous or puberulous to pubescent at first, glabrescent. Primary spines borne on the parent axis at a varying angle, (1–)2–5(–10) cm above the axil, 2–11 cm long, 2–5 mm diameter at the base, subulate, smooth or sometimes shallowly grooved, green or yellowish-green, spinuliferous with some branch-spines; spinules 0.5–2 cm long grading to branch-spines. Leaves on the stems and spines, sessile or with a petiole up to 4 mm long; stipules 1–2(–3) mm long, triangular, puberulous, often persistent; leaflets sessile or sub-sessile, orbicular to broadly obovate or obovate-elliptic, 0.7–6.5 cm long, 0.8–4.8 cm wide, coriaceous, frequently concave and undulate, apex rounded or sometimes emarginate to truncate and abruptly apiculate, base rounded or broadly cuneate, glabrous or puberulous to pubescent, eventually glabrescent, rarely setulose; foliole linear, 0.7–4 mm long, sometimes caducous. Inflorescence on the stems and spines, a few to 12-flowered fascicle or sometimes clustered on a short peduncle up to 2.8 cm long, axillary or rarely terminal on a leafy shoot; pedicels 0.2–1.6 cm long. Flowers 4-merous; sepals ovate, 3–5.5 mm long, 3 mm wide, acute to acuminate, sparsely pubescent outside, the glabrous margin narrow; petals yellowish green or olive-green, obovate-elliptic, 3–6.5 mm long, 1.8–2.5 mm wide, acute or obtuse and sometimes irregular at the apex, narrowing to the base, glabrous within; stamens 8, spreading-erect; anthers 1.5–2 mm long, 0.5–1 mm wide; ovary 1–1.5 mm high, glabrous or densely pubescent to pilose; style 1–2 mm long. Fruit, not elongating in early development, swelling at first proximally, with slower expansion of the glabrous style-base, eventually ripening orange-yellow, ovoid to broadly ellipsoid when mature, 2.5–3 cm long, 1.8–2.5 cm diameter, rounded at both ends, thin, hard, brittle and smooth on the outside, fibrous within enclosing a dense pale layer, the endocarp hard and oily, the seed becoming free inside the fruit.

Ovary densely pubescent to pilose; leaflets up to 6.5 cm long,
  persistent stiff hairs absent · · · · · · · · · · · · · · · · · · var. *rotundifolia*
Ovary glabrous or rarely with a few hairs when young; leaflets
  up to 2(–3) cm long, persistent, erect stiff hairs present var. *setulifera*

Fig. 2. *BALANITES PEDICELLARIS* subsp. *PEDICELLARIS*. **1**, habit showing flowers and fruits on spines, × 0.5; *B. ROTUNDIFOLIA* var. *ROTUNDIFOLIA*. **2**, habit, × 0.5; **3**, habit, × 0.4; **4**, inflorescence, × 3. 1 from *Semsei* 3090; 2 from *Carr* 655, 3, 4 from *Bally & Smith* 14903. Drawn by Margaret Tebbs.

var. **rotundifolia**; Sands in K.B. 56: 95, t. 14/a–m, p, map 9 (2001)

Leaflets up to 6.5 cm long, puberulous to pubescent and eventually glabrescent; pedicels 0.2–1.1 cm long, pubescent; ovary densely pubescent to pilose. Fig. 2/2–4.

UGANDA. Karamoja District: Lorengidwat, 1936, *Eggeling* E2846! & Moroto, Kasemen Estate, Dec. 1971, *J. Wilson* 2154! & Koputh, no date, *Brasnett* 178!
KENYA. Northern Frontier District: Tagaba, Jan. 1949, *Dale* K731!; Turkana District: 9.5 km N of Lodwar, 25 March 1954, *Hemming* 246!; Lamu District: Lunghi Forest Reserve, 1 Dec. 1988, *Luke & Robertson* 1536!
DISTR. **U** 1; **K** 1, 2, 7; Sudan, Ethiopia, Eritrea, Djibouti, Somalia; Yemen (Aden)
HAB. Deciduous bushland, wooded grassland; often on lava; 0–800 m

SYN. *Agialida rotundifolia* Tiegh. in Ann. Sci. Nat. sér. 9, 4: 230 (1906) & in F.R. 7: 119 (1909)
  Balanites orbicularis Sprague in K.B. 1908: 57 (1908); I.T.U.: 230 (1940); E.P.A.: 366 (1956); K.T.S.: 534 (1961). Types: Somaliland, *Drake-Brockman* 336 & 337 (K!, syn.)
  B. sp., Drake-Brockman, Brit. Somalil.: 306 (1912)
  B. patriziana Lusina in Ann. Bot. (Rome) 20: 138 (& 137) (1933); E.P.A.: 366 (1956); Cufod. in Senckenberg. Biol. 39: 303 (1958). Type: Eritrea, Dancalia Airori, wadi Gaarre, *Patrizi-Montoro* s.n. (RO, holo.)
  [*B. aegyptiaca sensu* Cufod. in E.P.A.: 365 (1956) pro parte (quod syn. *B. rotundifolia* (Tiegh.) Blatt.), *non* (L.) Delile]
  B. gillettii Cufod. in Senckenberg. Biol. 39: 302, t. 37/3–4 (1958); K.T.S.: 533 (1961). Type: Ethiopia, Burgi, valley east of Amaro Mts, *Gillett* 15069 (W, holo., EA!, FT!, K!, PRE!, iso.)
  B. gillettii Cufod. var. *renifolia* Cufod. in Senckenberg. Biol. 39: 303, t. 38/5 (1958). Type: Ethiopia, Fora Dara, *Kuls* 385 (W/WU?, holo.)

var. **setulifera** *Sands* in K.B. 56: 99, t. 14r, map 9 (2001). Type: Kenya, Tana River District, 72 km WSW of Galole, *Buechner* 17 (K!, holo.)

Leaflets with short stiff hairs which persist mainly on the lower surface. Ovary glabrous.

KENYA. Teita District: Tsavo National Park East, Voi Gate–Sobo Road turn off, km 45, 27 Dec. 1966, *Greenway & Kanuri* 12889!; Tana River District: 72 km WSW of Galole, 18 Feb. 1965, *Buechner* 17! & Galole, Nov. 1964, *Makin* in *E.A.* 13045!
DISTR. **K** 7; not known elsewhere
HAB. Deciduous bushland; 50–400 m

SYN. [*B. orbicularis* sensu Greenway, J. E. Africa Nat. Hist. Soc. Natl. Mus. 27: 189 (1969) pro parte quoad *Greenway & Kanuri* 12889]

6. **B. glabra** *Mildbr. & Schltr.* in E.J. 51: 163, fig. 1A/a–d (1913); V.E. 3: 743, fig. 348a (1915); E.& P. Pf. 19a: 182, fig. 87a (1931); T.T.C.L.: 571 (1949); K.T.S.: 534 (1956) pro parte excl. loc. Turkana; Blundell, Wild Fl. E Afr.: t. 15 (1987); Sands in Fl. Ethiop. 3: 434, fig. 121.1/7–9 (1990); K.T.S.L.: 378, map (1994); Sands & Thulin in Fl. Somal. 2: 169, fig. 111/g–i (1999); Sands in K.B. 56: 99, t. 15, map 10 (2001). Type: Tanzania, between Meandet and Kitumbini (District unclear), *Uhlig & Winter* 220 (B†, holo.); near Engare Nairobi, Moshi District, *Greenway & Kanuri* 12451 (K!, neo., BR!, EA, isoneo., chosen by Sands)

Very spiny evergreen shrub or small tree up to 9 m high, rarely taller, ± erect, sometimes drooping, rarely 'creeping' (*Napier* 2354) or subscandent; bark grey or grey-green, rough, corky, corrugated and fissured; branches smooth, yellowish green to dark green; branchlets and immature spines green, at first appressed-puberulous, glabrescent, the second-year stems very sparsely hairy, sometimes glaucous. Spines borne at right-angles to the parent stem 5–20(–30) mm above the axil, 3–13 cm long, 2–3.5 mm diameter, sometimes bearing branch-spines, very stout, terete, yellowish-green, the spine-tip often orange-brown; spinules absent. Scale-leaves 2 mm long, 1 mm wide, triangular, acute, scarious, sometimes persistent. Leaves infrequent, subsessile or with a petiole up to 1–2.5 mm long; stipules 1 mm long, very early caducous; leaflets elliptic, obovate or obovate-spathulate, sessile, 2.2–6.2 cm long, 1.2–3.8 cm wide, apex rounded, obtuse or acute, sometimes minutely apiculate, base cuneate, thinly coriaceous, glabrous above, often very sparsely appressed puberulous below; foliole 1–2 mm long, linear, rarely foliar and to 1.7 cm long, 0.4 cm wide. Inflorescence generally on the spines, but sometimes on the parent axis, fasciculate; pedicels 3.5–10 mm long, densely white-tomentellous. Flowers 4 (rarely 5)-merous, arising from a small swollen tomentellous cushion, sometimes scented; sepals ovate, 4.5–5 mm long, 2–2.5 mm wide, acute, caducous, shortly puberulous outside; petals yellowish green, cream or white, rarely orange, narrowly obovate-oblong, 6–7 mm long, 2–2.5 mm wide, acute, gradually narrowing to the base, glabrous within; stamens 8; anthers 2 mm long, 1 mm wide; ovary 1 mm high, densely silky hairy initially in bud, usually very early-glabrescent; style 1–1.5 mm long. Fruit not markedly elongating in early development, at first ellipsoidal and pointed at both ends, eventually ovoid, ripening yellow to orange or pale red, to 3.5 cm long, to 2 cm diameter, smooth; seed in a hard endocarp enclosed in a thin outer layer.

KENYA. Meru District: Isiolo, 3 July 1960, *Paulo* 496!; Machakos District: footslopes of Lukenya plateau, 28 Aug. 1959, *Verdcourt* 2359!; Masai District: Magadi road km 56, 10 March 1951, *Greenway* 8505!
TANZANIA. Mwanza District: Ilumia, Massanza I., 10 Oct. 1952, *Tanner* 1051!; Masai District: Great Ardai Plain, Monduli Mt, 9 July 1943, *Greenway* 6752!; Pare District: Kihurea-Dungu, 9 Sept. 1935, *Greenway* 4070!
DISTR. **K** 3, 4, 6; **T** 1–3; Ethiopia, Somalia
HAB. Deciduous bushland, scattered tree grassland, wooded grassland, thicket; 800–1700 m

Syn. [*B. aegyptiaca* sensu Drake-Brockman, Brit. Somalil.: 306 (1912); Burger, Fam. Fl. Pl. Ethiopia (Exp. Stat. Bull. No. 45): 163, fig. 25, 2 (1967), pro parte, *non* (L.) Delile]

NOTE. According to K.T.S. the species occurs in the Turkana region of Kenya, but no collections have been seen.

7. **B. 'Omo valley'**; Sands in K.B. 56: 105, t 2k, 3k/i & ii, map 10 (2001)

A prostrate to semi-erect shrublet to 55 cm high, stem up to 3 mm diameter, the whole plant pilose, densely so at first and only the two-year-old stem glabrous. Spines borne at a wide angle to the parent stem, 1–6 mm the axil, 0.5–1.1 cm long, 0.5–1 mm diameter at the base, very slender, often slightly bent at the nodes, gradually tapering to a fine point, most bearing (1–)2 spinules 0.5–2 mm long at right angles. Scale-leaves subtending the spinules, 0.25–0.5 mm long, linear, tardily caducous. Leaves all borne on the stem, sessile or subsessile; stipules 0.5–1 mm long, filiform, a few persistent for some time; leaflets ovate-elliptic, sessile or subsessile, 1.2–2.7 cm long, 0.3–1.4 cm wide, thin-textured, apex acute, mostly apiculate, base cuneate; foliole 1.5–3 mm long, filiform. Flowers and fruits not seen.

Although this taxon has not been found in Kenya, it was collected just over the Kenya–Ethiopia border in the lower Omo valley, and may occur in Kenya as well.

DISTR. Ethiopia. Known only from Omo R not far from Kelam (*Carr* 859, EA)
HAB. The precise habitat is not known but probably riverine woodland

INDEX TO BALANITACEAE

Agialida abyssinica Tiegh., 7
Agialida aegyptiaca (L.) Adans., 7
Agialida cuneifolia Tiegh., 7
Agialida latifolia Tiegh., 7
Agialida membranacea Tiegh., 7
Agialida nigra Tiegh., 8
Agialida rotundifolia Tiegh., 11
Agialida schimperi Tiegh., 8

Balanites, 1
Balanites aegyptiaca (*L.*) *Delile*, 6, 1, 8, 11, 13
   var. **aegyptiaca**, 6
   var. *ferox* (Poir.) DC., 8
   var. *pallida* Sands, 8
   var. **quarrei** (*De Wild.*) *G. Gilbert*, 8
   var. **tomentosa** (*Mildbr. & Schltr.*) *Sands*, 8
Balanites aegyptiaca auct., 11, 13
Balanites dawei Sprague, 5
Balanites fischeri Mildbr. & Schltr., 8
Balanites gillettii Cufod., 11
   var. *renifolia* Cufod., 11
Balanites glabra *Mildbr. & Schltr.*, 12
Balanites horrida Mildbr. & Schltr., 9
Balanites latifolia (Tiegh.) Chiov., 8
Balanites maughamii *Sprague*, 4
   subsp. **acuta** Sands, 5
   subsp. **maughamii**, 5
Balanites 'Omo Valley', 13

Balanites orbicularis Sprague, 11
Balanites orbicularis auct., 12
Balanites patriziana Lusina, 11
Balanites pedicellaris *Mildbr. & Schltr.*, 8
   subsp. **pedicellaris**, 9
   subsp. **somalensis** (*Mildbr. & Schltr.*) *Sands*, 10
Balanites quarrei De Wild., 8
Balanites rotundifolia (*Tiegh.*) *Blatt.*, 10, 9
   var. **rotundifolia**, 11
   var. **setulifera** *Sands*, 12, 9
Balanites somalensis Mildbr. & Schltr., 10
   var. *cinereo-corticata* Chiov., 10
Balanites suckertii Chiov., 8
Balanites tomentosa Mildbr. & Schltr., 8
Balanites wilsoniana *Dawe & Sprague*, 3, 5
   var. glabripetala Sands, 4
   var. mayumbensis (Exell) Sands, 4
   var. **wilsoniana**, 4
Balanites wilsoniana auct., 4, 5
Balanites sp. of Drake-Brockman, 11
Balanites sp. of Oliv., 5
Balanites sp. 1 of White, 5

Trachylobium mossambicensis Klotzsch, 5
Trachylobium mossambicensis auct., 5

Ximenia aegyptiaca L., 7

No new names validated in this part

GEOGRAPHICAL DIVISIONS OF THE FLORA

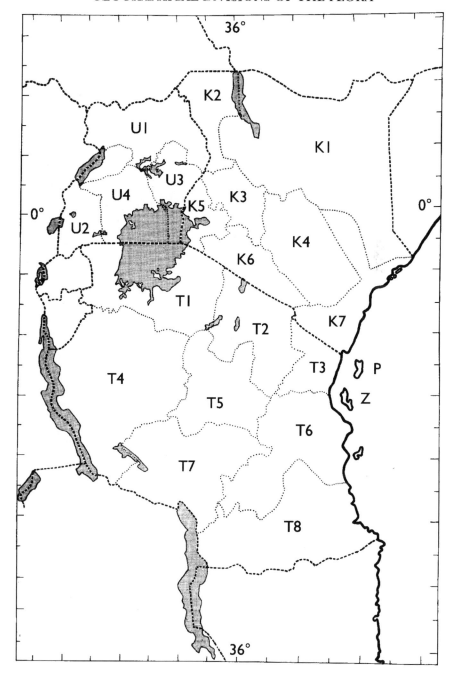

For Product Safety Concerns and Information please contact our EU representative GPSR@taylorandfrancis.com Taylor & Francis Verlag GmbH, Kaufingerstraße 24, 80331 München, Germany

Printed and bound by CPI Group (UK) Ltd, Croydon, CR0 4YY
01/05/2025
01858323-0001